I0396821

PREPPER SURVIVALIST HANDBOOK

EMERGENCY HOMESTEADING AND SURVIVAL GUIDE

NIKOLA MCKIMMY

Danny Gansneder

DISCLAIMER

ISBN-13:978-1512339987
ISBN-10:1512339989

CONTENTS

INTRODUCTION

If you are thinking about living off the grid, then you have taken an important first step in being prepared when an emergency strikes. Notice, the words when an emergency strikes, because one will surely strike sometime soon. It can be a manmade disaster like biological warfare, nuclear holocaust or overpopulation. Alternatively, it could be a major snowstorm, hurricane or some other natural disaster. Regardless, you need to be prepared to live off the grid.

When preparing, you will need to find the right answers for yourself and your loved ones. For many people, that means giving up living in a big city and moving to a rural setting. For others, it means being prepared to live in the city even when shopping is not available.

Additionally, you will need to consider what you plan to eat. Many people choose to start saving seed in preparation for that day. Others choose to stockpile food. Others choose to raise livestock. Still others choose to can their food.

You will also need to know how to prepare food if you have no electricity. For some, this is as simple as installing a large propane tank on their property. Others, choose to burn wood or other materials to cook and keep warm.

The correct answers to all these questions and more depends on your individual circumstances. It is important to think about the answers and start preparing now. That is why you need this book.

CHAPTER 1 WHERE TO LIVE

During the Cold War, United States Congress and their key staff members knew that they could head to the Greenbrier

Hotel in White Sulphur Springs, West Virginia, for safe shelter. The underground shelter contained everything needed by these people to continue their own existence and help protect the people of the United States.

Today, those same people can enter a tunnel that leads 140 feet underground. They would then climb on special vehicles that currently will haul them 6.5 miles to the Raven Rock Mountain Complex. This complex has a special ventilation system and is set to protect these government officials with chemical, biological, and radiation filters. The complex, deep underground in the middle of a mountain, even has a freshwater reservoir. Yet, the government is not finished. They are still digging the tunnel even deeper, and the new tunnel will be longer and steeper than the current one.

If the United States government is spending so much money preparing for a disaster, then you should be too. Unfortunately, depending on the government may not be an option, as they have already failed at the task. For example, as Hurricane Sandy battered Staten Island the Federal Emergency Management Association chose to close all of their locations. The New York National Guard followed quickly. Therefore, residents were left to fend for themselves.

SHELTER IN PLACE

The first line of defense should be a place to shelter in place in case of chemical, radiological or biological contaminants. Before you need to shelter in place, choose a room that is above ground level and has very few openings. Take plastic that is at least 4 millimeters thick and cut to fit each opening. Label each opening for the hole it is designed to fit. Equip the area with a landline phone and have a battery operated or hand cranked radio in the room. When it is necessary to shelter in place, then attach the plastic to the openings. Turn off all fans along with heating and air conditioning systems. You should be prepared to do this at home, school and work. Use the radio to learn what is going on and what government leaders think you should do.

PANIC ROOMS

This plan is the bare necessity, and many people choose to do much more to protect themselves and their loved ones. For example, some people choose to build a panic room in their homes. These rooms contain extremely thick concrete walls and only one door. That door is a solid core wooden or steel door that opens outward and hangs from a steel door trim. Many people install a keyless deadbolt on these rooms. The air that is in the room is carefully controlled. The room should contain lighting that is off the grid. Often these rooms are hidden. If you are planning on building a panic room for possible long term living, then it needs to have a minimum of 20 square feet per person. You also need to think about storing food in the room. Many people even store personal protective equipment in their rooms.

Other people choose to build a separate panic room that is not in the main building. When constructing these rooms, make sure that you can easily access it from the main house. Perhaps, through an underground tunnel.

UNDERGROUND SHELTERS

If you are going to build an underground shelter, then a practical way to do it is to use a shipping container that is in great condition. Dig a hole in the ground that is 24 inches deeper than the shipping container. Pour concrete at the bottom of the hole and build concrete walls to reinforce the outside. Now, coat the outside of the shipping container with stucco so that it is watertight. Insert the shipping container into the hole and reinforce the top with more dirt and concrete. Make sure to install air ventilation and lighting.

REMOTE LOCATIONS

Other people choose to buy land in a remote area. If you are not ready to live on this land full-time, then consider how you will get to the location if an emergency occurs.

Finally, some people choose to leave urban areas altogether and head to remote locations often referred to as bug-out locations. When choosing a bug-out location, your first consideration should be finding a location that offers water. Then, look for a location where you can grow a crop and livestock. Consider a location where there is abundant firewood, and with natural resources including wild animals. Finally, consider the weather and how it will affect your ability to live in the bug-out location long-term.

CHAPTER 2: WHAT TO EAT

Stop and imagine a world where there are no grocery stores to run to when you run out of food. In fact, imagine a world where money has no value because there is nothing available to buy. Now, imagine if you were in that situation for days, weeks, or maybe even months. If this thought scares you, then you need to learn to prepare your own food.

In fact, that scenario is what many governments around the world are trying to prepare for today. For example, the Alaska state government is currently stockpiling enough food to feed 40,000 people for a week in warehouses near their military bases. Numerous reports indicate that the United States government is stockpiling food and has been since 2008. If you are wise, and we know that you are, then you may want to start stockpiling food too.

72-HOUR EMERGENCY KITS

The first step that everyone should take is to prepare a 72-hour emergency kit. Even if you do not believe that the world as you know it will break down, this kit is essential in the case of storms and other natural catastrophes. As the name implies, it is meant to last 72 hours. Your kit should be with you at all times, or you will need to prepare a separate one for home, school, work and your vehicle.

SHORT TERM EMERGENCY WATER

The first thing that you need for your 72 hour emergency kit is water. If the electricity is not working, then water pumps and water treatment facilities cannot operate. There should be a minimum of three gallons of water per person in each emergency kit. The weight of water may prohibit carrying kits containing water a long way, so consider including a way to purify water.

You can accomplish simple purification in three different ways. First, you can boil the water for at least four minutes. Secondly, you can add 16 drops of unscented bleach for every gallon of water. Combine the water and the bleach and let sit for at least 30 minutes before using. You should smell bleach after the 30 minutes is over. If not, then repeat up to two more times. If you still do not smell bleach, then discard the water. Finally, you can distill water.

SHORT TERM EMERGENCY FOOD

The second thing that you need in your 72-hour emergency kit is food. While many experts recommend that you keep survival bars in your bug out bag, you need to consider other alternatives. For example, dehydrated fruits and vegetables are calorie dense, and you will be operating in an environment that is full of stress and requires lots of energy. Secondly, consider including whole grain crackers, as they will keep you full for a long time without adding much weight to your bag. In addition, add some packages of non-salted nuts as they serve as protein. Make sure that they are without salt, because salt will dehydrate you. Finally, add some canned meats.

CHAPTER 3 STOCKPILING FOOD

You will also need a year's worth of food stockpiled even if you plan to start growing your own food and raising livestock. In order to prepare for each person over seven, you will need:

Amount	Food
150 pounds	Wheat
25 pounds	Flour
25 pounds	Cornmeal
25 pounds	Oats
50 pounds	Rice
25 pounds	Pasta
4 pounds	Shortening
2 gallons	Vegetable oil
2 quarts	Mayonnaise
1 quart	Salad dressing
4 pounds	Peanut butter
30 pounds	Dry beans
5 pounds	Lima beans
10 pounds	Soy beans

5 pounds	Split peas
5 pounds	Lentils
5 pounds	Dry soup mix
3 pounds	Honey
40 pounds	Sugar
3 pounds	Brown sugar
1 pound	Molasses
3pounds	Corn Syrup
3 pounds	Jam
6 pounds	Powdered Fruit Drink
1 pound	Flavored Gelatin
60 pounds	Dry milk
1.25 gallons	Evaporated milk
1 pound	Baking Powder
1 pound	Baking soda
.5 pounds	yeast
.5 pounds	salt
.5 gallons	vinegar

If you are likely to have children between the ages of two and seven, then you can cut these amounts in half for them. If you are likely to have a baby with you, then remember to store formula and a way to prepare baby food as the child grows.

FOOD STORAGE FOR LONG-TERM SURVIVAL

Consider how you will store the food to protect it from the environment and pests. The most effective way to do this is with 30-gallon food-grade steel drums, although 5-gallon food grade plastic buckets can be used. You will also need 7 millimeter thick Mylar bags, oxygen absorber packets, an iron, paper file labels, and a carpenter's level. Start by attaching two paper file labels to each bag that you will be filling and labeling them with the contents you plan on putting in the bag. Then, fill the bag with food making sure that the sealing surface stays clean. Add the oxygen absorber packet. Now, lay the bag on top of the carpenter's level. Using a hot iron without any steam, seal the bag ¾ of the way shut. The carpenter's level reflects the heat back to the bag so that you can seal both sides at once. Then, squeeze out as much air as you can. Return the bag to the top of the carpenter's level and seal the bag the rest of the way shut. Insert the bag into your storage container and seal tightly. Label the outside of the bucket with its contents.

CHAPTER 4 GARDENING FOR LONG-TERM SURVIVAL

Even though the beans and peas in your year's supply of food will help feed your family, it is not enough. You need to be growing food. Start now, so that if the need arises, then you are already an expert. That way your family is not relying on your first crop.

OPEN-POLLINATED SEEDS

Grow open-pollinated seeds so that you can save the seeds. In order to have enough food to feed your family from your garden, you will need the following for each person:

WHAT TO GROW

Type of Fruit or Vegetable	Number to Plant
Asparagus	30
Beets	4
Bush Beans	12

Pole Beans	2
Cabbage	4
Carrots	6
Cauliflower	2
Corn	4
Cucumbers	1
Greens (Spinach)	3
Kale	2
Onions	2
Peas	3
Peppers (Sweet)	1
Peppers (Hot)	1
Summer Squash	2
Tomato	2
Turnips	3

You will want to adjust this list to fit the tastes of your family. By planting this number of plants, you should have enough to eat for one year and can enough for two years. Save the seeds from the best plants and repeat the following growing season. If you need detailed gardening instructions, then be sure to check out The Seed Saving Guide: Beginner's Guide to Growing and Saving Seed along with Organic Gardening: Starting Your Own Healthy and Natural Garden.

CHAPTER 5 CANNING FOR LONG-TERM SURVIVAL

Canning is economical and smart way to prepare for a food shortage. You can store properly canned food for 18 months. Here are some recipes that you may want to follow:

CANNED BELL PEPPERS

1 pound peppers yields 1 pint canned peppers

½ teaspoon salt per pint

Instructions:

1. Core and seed peppers.
2. Place peppers in 400 degree Fahrenheit oven for seven minutes until skin blisters
3. Allow peppers to cool
4. Place peppers on cookie sheet and cover with damp dish towel.
5. Wait 10 minutes
6. Flatten peppers.
7. Add ½ teaspoon salt to each pint jar.
8. Fill jar loosely with peppers
9. Bring water to boil
10. Carefully pour water over peppers leaving 1 inch at top of jar.
11. Screw on lids.
12. Process for 35 minutes in a dial-gauge canner at 11 pounds pressure for altitudes up to 2,000 feet; 12 pounds pressure for altitudes between 2,001 feet and 4,000 feet, 13 pounds pressure for altitudes between 4,001 and 6,000 feet or 14 pounds pressure for altitudes between 6,001 and 8,000 feet. If using a weighted-gauge pressure canner, then process at 10 pounds pressure for altitudes less than 1,000 feet and 15 pounds pressure for altitudes over 1,000 feet.

CANNING GREENS

Many people may wonder if they can preserve greens such as kale, spinach, mustard greens and collards by canning. The answer is a resounding yes. Here is how:

1. Chop greens into bite size pieces
2. Rinse greens thoroughly

3. Fill stockpot with greens
4. Fill stockpot with water
5. Cook over medium-high heat until greens begin to wilt
6. Meanwhile, place pint jars and lids in another pan and fill with water. Bring to a boil.
7. Remove one jar at a time and fill with greens
8. Add enough water from cooking pot to cover greens leaving 1 inch headspace
9. Add ½ teaspoon canning salt to each pint
10. Wipe jar rim and put on lids
11. Process for 70 minutes at 10 pounds of pressure

PICKLING VEGETABLES

In addition, think about adding pickled vegetables to your long-term food supply. While you can pickle individual vegetables, combining several different vegetables allows you to can food that you might not have enough of otherwise. Here is one recipe but feel free to make it your own using what you have in the garden. Here is the recipe:

8 pounds pickling cucumbers about 5 inches long

4 pounds small onions

8 cups celery

4 cups carrots

4 cups bell pepper

4 cups cauliflower

10 cups white vinegar

½ cup prepared mustard

1 cup canning salt

7 cups sugars

6 tablespoons celery seed

4 tablespoons mustard seed

1 tablespoon whole cloves

1 tablespoon ground turmeric

Instructions:

1. Cut vegetables into bite size pieces and place in a large baking dish
2. Cover with two inches of crushed ice
3. In a 16-qt. pan, combine vinegar and prepared mustard
4. Add spices to vinegar and mustard mixture
5. Bring mixture to a boil
6. Drain vegetables and add them to the mixture
7. Cover and return to a boil
8. Drain vegetables, but save the mixture
9. Fill jars with vegetables leaving ½ inch empty at top
10. Add mixture making sure to still leave ½ inch empty at top
11. Screw on lids
12. Process quarts in boiling-water canner for 10 minutes for altitudes less than 1,000 feet, 15 minutes for altitudes between 1,001 and 6,000 feet and 20 minutes for altitudes higher than 6,000 feet

CHAPTER 6 RAISING LIVESTOCK

In addition to canned food, many people choose to raise animals and butcher them for the meat.

The first animal that many people choose is the chicken. When choosing chickens to raise for long-term survival, consider your environment. Here are some suggestions:

- Aseels hens have a natural mothering instinct.
- Brahmas are serene and adjust to their environment easily. They also tolerate cold temperatures well.
- Buckeyes tolerate cold temperatures well even as babies. The meat on these chickens is extremely flavorful.
- California Whites lay many eggs each year
- Chanteclers do well in cold temperatures.
- Cherry Eggers start laying eggs when only about 17 weeks old and are very productive.
- Cochins do not fight easily.
- Cochins hens are outstanding mothers who will even raise chicks that are not theirs.
- Cubalayas love when it is warm outside.
- Dorkings produce outstanding meat that is extremely flavorful.
- Faverolles are tranquil.
- Golden Comets are productive egg layers who start laying eggs when they are about 17 weeks old.
- Houdans do well as free-range chickens.
- Hy-line Browns are great egg layers
- Indian Rivers start laying eggs when about 17 weeks old and lay an egg almost every day.
- Javas can live even as babies when it is cold outside or in hot temperatures.
- La Fleche produces outstanding meat.
- Langshans do well as free-range chickens.

- Malays tolerate heat well.
- Orpingtons are easy to control.
- Pearl Leghorns start laying eggs when about 17 weeks old and do well as free-range chickens.
- Polish do well as free-range chickens.
- Silkies are calm, and the hens make great mothers.
- Sumatras does well in warmer temperatures.

Make sure to consider how you will keep the chickens where they belong and away from wild animals. Generally, you will need four square feet of space in the coop for each chicken and 10 square feet of space in the chicken run.

Remember that in emergency situations, chickens may be very vulnerable to attack from others for their meat and eggs. The right number of chickens depends on the breed and your family. Generally, you will need two hens for each family member plus one or two extra.

RABBITS

Another animal that is great for raising for long-term survival is rabbits. While there are many cute breeds of rabbits for long-term survival you need to consider meat producing rabbits. The most popular of these are Californians and New Zealands. Most people start off with two breeding trio sets with each consisting of one buck and two does. While rabbits can have up to 14 babies per litter, the average is six. It takes about three months to raise a rabbit to the fryer stage where most people butcher the rabbit for meat. Each rabbit needs about 2.5 square feet of cage space. When getting ready to butcher, consider which rabbits you want to mate. Be very careful to not breed the same lines too often. Therefore, after the first mating, you may want to secure another buck that is unrelated to your current stock.

PIGS

Pigs are another animal that people can easily keep for long-term survival. One of the reasons is that they can be left to forage for food, you can feed them vegetable scraps, or you can raise corn to feed them. Most pigs are killed for food when they weigh about 250 pounds. This yields about 185 pounds of meat including about 55 pounds of ham and shoulder and 40 pounds of bacon and loin. Many raisers prefer crossbred pigs for meat. The most common crossbreeds are Duroc because of the meat quality crossed with either Chester White or Berkshire for hardiness. Raising pigs will require one boar and at least two sows for starters. Be extremely careful around your pigs as boars and sows with piglets can be extremely aggressive.

GOATS

Goats are another excellent animal for long-term survival. The first reason is that they are very easy to raise and will eat almost anything. Secondly, producers raise them for their milk. Unlike cows, they take up less space and continue to provide milk for up to two years following the birth of their baby. Goat's milk is easily consumed by humans and by other animals. Goats can also be taught to pull small wagons that can be a real asset when working on a farm. If you have a very small space, then consider Nigerian Dwarfs that stand only 18 inches tall when mature. This breed averages one or two quarts of milk that averages 7 percent butterfat each day. Nubian, Alpine, and LaMancha goats each produce about one gallon of milk per day that contains about four percent butterfat.

CHAPTER 7 LONG TERM MEAT STORAGE

After you have raised the animal and butchered it, then you will need to keep the meat. Remember that you may have no electricity, so you may need to choose solutions besides putting it in the freezer or refrigerator.

CANNING MEAT

Canning meat is one solution. Start by allowing the meat to get cold. Then, soak the meat in a solution of 1 tablespoon salt to one-quart water for about seven hours. Then, cut the meat up so that it fits in your jars leaving 1 ½ inch empty headroom at the top of the jar. Insert meat into jars and add 1 teaspoon salt per quart. Process the meat in a dial-gauge pressure canner for 90 minutes for quarts. If you are at an altitude less than 2,000 feet, then you need 11 pounds per square inch pressure If you are at altitude between 2,001 and 4,000 square feet, then you need 12 pounds per square inch If you are between 4,0001 and 6,000 square feet, then you need 13 pounds per square inch. Above that you need 14 pounds per square inch. Note that rabbit and poultry can be cooked until it is 70 percent done before canning. If you are canning raw meat, over time, the meat makes its own juices.

SMOKING MEAT

Smoking meat is another popular way of preserving it. You can make a simple smoker out of materials that you will likely have on hand. Start by building a teepee out of three poles. Then, surround the teepee with ponchos on all sides. Start a small hardwood fire at the base of the teepee. Cut the meat into small strips and hang over the fire using fishing line. Let the meat smoke for two days. Smoking meat preserves the meat for up to a month.

DRYING MEAT

Drying meat is another popular way to preserve it. Simply cut the meat into thin strips and soak them in a 14 percent salt solution for five minutes. Drain the meat and hang from long poles where the sun can reach the meat. Do not allow the meat strips to touch each other. Cover the meat with ponchos or tarps to keep off the flies, though the salt does this naturally. The meat should continue to hang until it has lost 65 percent of its original weight. The salt acts as a natural way to kill bacteria that might attack your meat.

BRINING MEAT

Outside of keeping the meat alive on the animal, the best way to preserve meat is by brining it. Start by cutting the meat into slabs that are manageable in size. Then, combine 1 pound pickling salt and ½ cup brown sugar in three cups of water. Stir until completely dissolved. Place the meat in a glass pan and pour the mixture over the meat ensuring that you completely cover the meat with the mixture. Let set for one week in an area with a temperature of about 36 degrees Fahrenheit. Remove the meat from the brine. Stir the brine and then replace the meat. Continue the process for four more weeks. If you notice that the brine become stringy, remove the brine and wash the pan. Then, wash each piece of meat. Start again with a fresh batch of brine. Again, the salt acts to kill bacteria that might attack the meat. The meat is best used in stews as it will taste very salty.

CHAPTER 8 LONG-TERM ENERGY NEEDS

If disaster strikes, then you need to be able to produce some energy on your own. There is no system that can economically power all your household appliances at the moment, but you do have alternatives that will generate a limited amount of power.

SOLAR

The first source you may want to consider is solar energy. The major advantage of solar power is that the sun gives off enough energy to power many of your home's appliances. The cost of solar panels or supplies to build your own can be astronomical. In addition, most systems are mounted on the roof which must be strong enough to withstand the system's weight. The solar power system must ensure as much sun as possible. Current systems are extremely large so having enough room may be a problem in some locations.

WIND

A second source of possible energy is the wind. Like solar polar, wind energy can be very costly to produce because the cost of turbines big enough to power your home can be enormous. In addition, most wind turbines take a battery that needs replacing on a regular basis. The wind must blow regularly for this system to be effective. Finally, wind turbines require a large area to generate enough power for your entire home, so may not be suitable for urban areas.

GEOTHERMAL

A third alternative is geothermal energy. Like wind and solar energy, startup costs can be enormous. This form of energy relies on harnessing the heat from hot rocks below the earth's surface. Therefore, it is only suitable for certain areas. Over time, this energy source can be used up. Geothermal energy can produce toxic gasses that must be contained.

HYDROPOWER

Hydropower can be used in some areas. This form of energy relies on water vapor to be captured, so it must be located near a reliable source of running water such as a large stream or river. As with other sources of energy, startup costs can be enormous.

BIOMASS

Biomass energy can be used. This process requires the combustion of wood, trash or other materials. It often produces greenhouse gases. In an environment where everything is precious, there may be a limited amount of things that you can burn over a long time. Biomass energy can be difficult for the home consumer to capture.

CHAPTER 9 PROTECTION

Remember that in a long-term survival situation you may have to depend on yourself for protection. While Hurricane Katrina may seem mild compared to some events that require long-term survival skills, during that storm 200 officers in New Orleans quit. Therefore, you need to realize that you need to be able to protect yourself.

SELF-DEFENSE

Every member of the family should receive training in some form of self-defense. The skills learned in classes not only help you develop the discipline needed for long-term survival, but also help you become physically fit. Some possible choices include jujutsu, judo, krav maga or aikido. While there are differences in technique, all depend on rendering the attacker defenseless by putting their body into unusual positions, stopping the air supply, and attacking vulnerable joints such as the elbow and knees.

SURVIVAL KNIVES

Preparing for survival may also mean getting a knife and knowing how to use it. Since you will need to use your knife for many purposes, you will want a fixed-blade knife that is about 10 inches long with a pointed tip. The knife should be a full tang to stand up better under rugged conditions. The back side of the knife should have a 90-degree grind. Finally, look for a knife that has a solid butt.

SURVIVAL GUNS

Chances are that you will also need a survival gun. Consider who is going to be using the gun before making a purchase as smaller frame people may feel more comfortable with a smaller gun. If your plan involves moving, then consider the weight of the firearm. Look for a gun that is easy to maintain and has a great track record for being reliable. During a survival situation, gunsmiths may be very hard to find. Make sure to stockpile ammunition for the gun.

CONCLUSION

Prepare now for long-term survival. Most educated people believe it is not a matter of if disaster will strike, but when. Therefore, the wise person will start preparing today.

The first step is to know where you will stay during an event that could last from several days to the rest of your life. Therefore, you need a 72-hour plan and a long-term plan. Short-term you may choose to shelter in place or a separate facility on your property. Long-term survival may depend on going to a separate location where you have more natural resources available.

You also need to start preparing food for your survival. Again, start by preparing a 72-hour kit with the food that you are most likely to use to supply energy to your body. Then, start saving a year's worth of food. Learn how to garden now as your family may depend on your skills later, so you do not want crop failure. Learn which crops grow best in your area, and how to save seeds from them. Then, focus on getting animals that will help you feed your family.

In a long-term survival situation, you will need to be able to produce some energy. You have several choices depending on your environment. If installed professionally, the startup costs can be enormous. Therefore, choose the best one for your environment and start building the system at home.

Finally, you will need to protect yourself. Everyone should know self-defense. You may also want to secure a great survival knife and a survival gun that fits you.

THANK YOU

I promote eBooks to my subscribers. You will be the first to know when new books are published. I publish books on a wide variety of topics including healthy natural foods, self-help, health and natural healing alternatives, gourmet recipes, desserts and pastries, just to name a few. CLICK HERE to start receiving your eBooks and look for promotional or contest offers, such as gift cards, free vouchers, electronics and more!

If you have truly found value in my publication please take a minute and rate my book, I'd be eternally grateful if you left a review. As an independent author I rely on reviews for my livelihood and it gives me great pleasure to see my work is appreciated. You can also visit and like our Facebook page by clicking here.